GEOLOGICAL SURVEY OF CALIFORNIA.

J. D. WHITNEY, STATE GEOLOGIST

MINING STATISTICS.

No. 1.

TABULAR STATEMENT OF THE CONDITION OF THE AURIFEROUS QUARTZ MINES AND MILLS

IN THAT PART OF MARIPOSA AND TUOLUMNE COUNTIES LYING BETWEEN THE MERCED AND STANISLAUS RIVERS, AUGUST—NOVEMBER, 1865.

BY A. RÉMOND.

PUBLISHED BY AUTHORITY OF THE LEGISLATURE OF CALIFORNIA.

APRIL, 1866.

Science

NOTE.—The following tabular statement contains the results of an examination of seventy-seven gold mines and sixty-five quartz mills, situated in the region between the Merced and Stanislaus Rivers, all of which were visited by Mr. Rémond during the months of August to November, 1865. Of the seventy-seven mines, fifty-six were being worked at that time; and of the sixty-five mills, twenty-three were running The mines are distributed through the different kinds of rocks as follows:

In Granite,	23
In Mica Slate,	23
Partly in Granite and partly in Mica Slate,	5
In Clay Slate,	3
In the Green Porphyritic Schists,	3
Along the Great Quartz Vein,	20

SHERMAN & CO., PRINTERS,
PHILADELPHIA.

Number, ...	*1.*	*2.*	*3.*	*4.*	*5.*	*6.*
Mine, ...	**French Mary.**	**Hope.**	**Victor.**	**Mount Hope.**	**Catherine.**	**Cranberry.**
Location, ...	Merced Mountain.	Merced Mountain.	Merced River.	Merced River.	Near Merced River.	Near Merced River.
County, ...	Mariposa.	Mariposa.	Mariposa.	Mariposa.	Mariposa.	Mariposa.
Date of visit, ...	July 30th, 1865.	July 30th.	August 9th.	August 9th.	August 9th.	August 10th.
Whether worked, . .	Not worked.	Not worked.	Worked.	Not worked.	Not worked.	Worked.
Vein inclosed in, . .	Porphyritic schists.	Porphyritic schists.	Granite.	Granite.	Granite.	Granite.
Strike, ...	N. 35° W.	N. 55° W.	E.–W.	?	N. 30° W.	N. 65° W.
Dip, ...	N.E., 45°.	N.E., 60°.	N., 30°–35°.	?	E.N.E., 40°.	N.N.E., 45°.
Average width, . . .	1 foot 6 inches.	1 foot.	1 foot.	?	1 foot 2 inches.	2 feet 6 inches.
Veinstone, ...	White Quartz.	White Quartz.	Quartz.	Quartz.	Quartz.	Ferruginous Quartz.
Sulphurets present, .	Copper, Iron, Lead.	Copper, Iron, Lead.	Iron, Zinc, Lead.	Iron, Zinc, Lead.	None seen.	Copper, Lead, Iron.
Yield of ore per ton, .	$12.00	$6.00.	$10 to $60.	?	$25 to $40.	$16 to $20.
Cost of extraction, . .	?	$3.00.	?	?	$14.00.	$3.50.
Transportation to mill,	?	$1.00.	?	?	$1.25.	$2.00.
Cost of treatment at mill,	?	$1.50.	?	?	$3.00.	$3.00.
Mill, ...	**None.**	**Brichman's.**	**Victor.**	**Mount Hope.**	**Catherine.**	**Yosemite.**
Location, ...	. . . .	Merced River.	Merced River.	Merced River.	Merced River.	Merced River.
Whether running, . .	. . . .	Not running.	Not running.	Not running.	Not running.	Running.
When erected, . . .	. . . .	1864.	December, 1862.	May, 1864.	January, 1863.	December, 1860.
Cost, ...	. . . .	$4000.	$3000.	$3500.	$3600.	$3000.
Power employed, . .	. . . .	Water.	Water.	Water.	Water.	Water.
Number of stamps, . .	. . . .	10.	5.	5.	3.	5.
Kind of amalgamating machinery,	. . . .	Copper plates.	Copper plates and Arrastra.	Copper plates.	Copper plates and 2 Arrastras.	Copper plates.
Cost of roads or trails,	. . . .	$500.	$100.	$100.	$1000.	$1000.
Cost of flumes & ditches,	. . . .	$500.	$1000.	$1000.	$600.	$1200.

Number,	*7.*	*8.*	*9.*	*10.*	*11.*	*12.*
Mine,	**Rutherford's.**	**Ferguson's.**	**Cedar.**	**Empire.**	**Mary Harrison.**	**Malvina.**
Location,	Merced River.	Merced River.	Near Bull Creek.	Marble Spring Gulch.	Near Coulterville.	Near Coulterville.
County,	Mariposa.	Mariposa.	Mariposa.	Mariposa.	Mariposa.	Mariposa.
Date of visit,	August 10th.	August 10th.	August 13th.	August 14th.	August 16th.	August 17th.
Whether worked,	Not worked.	Worked.	Worked.	Not worked.	Worked.	Worked.
Vein inclosed in,	Granite.	Mica Slate.	Granite & Mica Slate.	Mica Slate.	Clay Slate and Quartz	Clay Slate.
Strike,	N. 10° E.	N. 10° E.	N. and S.	N. 20° W.	N. 50° W.	N. 60° W.
Dip,	E., 65°.	W. 75°.	W., 55°–60°.	E.N.E., 50°.	N.E., 60°–65°.	N.N.E., 60°–65°.
Average width,	1 foot 6 inches.	8 inches.	1 foot.	1 foot 6 inches.	3 feet.	10 feet.
Veinstone,	White Quartz.	Quartz.	Ferruginous Quartz.	White Quartz.	Quartz, Limestone, Talc.	White Quartz.
Sulphurets present,	Iron, Zinc, Lead.	Arsenic, Iron, Lead, Zinc.	Iron.	Iron, Lead, Zinc.	Iron.	Iron, Copper.
Yield of ore per ton,	$25.00.	$20.00.	$40.00.	$40.00.	$8 to $20.	$8 to $30.
Cost of extraction,	?	?	$4.00.	?	$4.00.	$1.75.
Transportation to mill,	?	?	$2.50.	?	$1.00.	$0.87½.
Cost of treatment at mill,	?	?	$2.50.	?	$2.00.	$2.00.
Mill,	**(No. 6.)**	**Ferguson's.**	**Cedar.**	**Empire.**	**French Mill (old).**	**French Mill (new)**
Location,	. . . .	Merced River.	Bull Creek.	At Mine.	Maxwell Creek.	Maxwell Creek.
Whether running,	. . . .	Running.	Running.	Ruined.	Not running.	Running.
When erected,	. . . .	December, 1863.	October, 1865.	1851.	April, 1862.	September, 1864.
Cost,	. . . .	$6000.	$3000.	?	?	?
Power employed,	. . . .	Water.	Water.	. . . .	Steam, 20-horse.	Steam, 60-horse.
Number of stamps,	. . . .	8.	5.	. . . .	15.	35.
Kind of amalgamating machinery,	. . . .	Copper plates.	Copper plates.	. . . .	6 Hungarian pans.	14 Hungarian pans and 1 Iron Arrastra.
Cost of roads or trails,	. . . .	$500.	$90.00.	None.	$330.	$150.
Cost of flumes & ditches,	. . . .	$1000.	$150.	None.	None.	None.

Number,	*13.*	*14.*	*15.*	*16.*	*17.*	*18.*
Mine,	**Adelaide.**	**M'Alpine's.**	**Louisiana.**	**Schimer's.**	**Funk's.**	**Casabon's.**
Location,	3 m. from Coulterville.	Quartz Gulch.	2 m. from Bower Cave.	Near Louisiana.	Gentry's Gulch.	N. Fork, Merced.
County,	Mariposa.	Mariposa.	Mariposa.	Mariposa.	Mariposa.	Mariposa.
Date of visit,	August 20th.	August 22d.	September 1st.	September 1st.	September 2d.	September 2d.
Whether worked,	Worked.	Not worked.	Worked.	Worked.	Worked.	Worked.
Vein inclosed in,	Clay Slates.	Great Quartz Vein.	Mica Slate.	Mica Slate.	Mica Slate.	Mica Slate.
Strike,	N. 65° W.	N. 70° W.	W.N.W. and E.S.E.	E.S.E.–W.N.W.	E.–W.	E.S.E.–W.N.W.
Dip,	N.N.E., 55°.	N.N.E., 50°.	N.N.E., 50°–60°.	N.N.E., 60°.	N., 25°.	N.N.E., 60°–70°.
Average width,	2 feet.	1 foot.	1 foot 6 inches.	1 foot.	10–12 inches.	1 foot 6 inches.
Veinstone,	White Quartz.	Quartz.	Quartz.	Quartz.	White Quartz.	Quartz.
Sulphurets present,	Iron	None seen.	Iron, Lead, Zinc.	Iron, Lead, Zinc.	None seen.	None seen.
Yield of ore per ton,	$6.00.	$25 to $50.	$25.00.	$15.00.	?	$15.00.
Cost of extraction,	$1.00.	$4.00.	$4.00.	$3.00.	?	?
Transportation to mill,	$1.25.	$2.00.	$0.75.	None.	?	?
Cost of treatment at mill,	$0.75.	$2.00.	$2.00.	$2.00.	?	?
Mill,	**Crown Lead.**	**M'Alpine.**	**Louisiana.**	**Low Mill.**	**Funk's (2 mills).**	**Casabon's.**
Location,	Merced River.	Near Mine.	Near mine.	On Mine.	Near Mine.	Near Mine.
Whether running,	Not running.	Not running.	Running.	Not running.	Not running.	Not running.
When erected,	October, 1863.	1857.	March, 1863.	1862.	December, 1862.	January, 1862.
Cost,	$42,000.	$10,000.	$6000.	$6000.	$5700.	$6000.
Power employed,	Water and steam, 10-horse.	Water.	Steam, 6-horse.	Steam, 14-horse.	Water.	Water.
Number of stamps,	25.	8.	5.	8.	10.	8.
Kind of amalgamating machinery,	4 Patterson's pans and 2 Separators.	Copper plates, and 2 Amalgamating Pans.	1 Salmon's Amalgamator, and 1 Salmon's Separator.	Copper plates.	Copper plates.	Copper plates.
Cost of roads or trails,	$6500.	$100.	$50.00.	$4.00.	$100.	$100.
Cost of flumes & ditches,	$18,000.	$200.	None.	None.	$500.	$3000.

Number, ...	*19.*	*20.*	*21.*	*22.*	*23.*	*24.*
Mine, ...	**Goodwin's.**	**Derrick's.**	**Humbug.**	**Blue Ledge.**	**Heslep's.**	**App's.**
Location, ...	Near N. Fork, Merced.	Near N. Fork, Merced.	Near Bull Creek.	1 mile from Black's.	Quartz Mountain.	Quartz Mountain.
County, ...	Mariposa.	Mariposa.	Mariposa.	Mariposa.	Tuolumne.	Tuolumne.
Date of visit, ...	September 3d.	September 3d.	August 14th.	August 14th.	September 11th.	September 11th.
Whether worked, . .	Not worked.	Not worked.	Not worked.	Worked.	Worked.	Worked.
Vein inclosed in, . .	Mica Slate.	Mica Slate.	Granite.	Mica Slate.	Talcose Slate & Great Quartz vein.	Talcose Slate & Great Quartz vein.
Strike, ...	E.–W.	E.–W.	N.–S.	E. S. E.–W. N. W.	N.W.–S.E.	N. 50° W.
Dip, ...	S., 70°.	N., 25°.	?	N. N. E.	N.E., 80°.	N.E., 60°.
Average width, . . .	2 feet.	1 foot 6 inches.	?	?	10 feet.	6 feet.
Veinstone, ...	White Quartz.	Quartz.	Quartz.	Quartz, Limestone, Clay.	Slate and Quartz.	White Quartz.
Sulphurets present, .	Iron.	Copper, Zinc, Iron, Lead.	None seen.	None seen.	Iron.	?
Yield of ore per ton, .	$15.00.	$25.00.	$4.00.	$23.00.	$12.00.	$18.00.
Cost of extraction, . .	$2.00.	$12.00.	$1.50.	$5.00.	$2.50.	$3.00.
Transportation to mill,	$1.50.	$3.00.	$0.50.	$2.00.	$0.40.	$0.75.
Cost of treatment at mill,	$1.50.	$4.00.	$3.00.	$3.00.	$2.00.	$2.50.
Mill, ...	**Eclipse.**	**Derrick's.**	**Humbug.**	**Black's.**	**Heslep's.**	**App's.**
Location, ...	N. Fork, Merced.	N. Fork, Merced.	Bull Creek.	Bull Creek.	On Mine.	Wood's Creek.
Whether running, . .	Not running.	Not running.	Not running.	Not running.	Running.	Running.
When erected, . . .	1863.	July, 1859.	1858.	1864.	July, 1854.	1857.
Cost, ...	$4000.	$3000.	$2300.	$1500.	$12,000.	$4500.
Power employed, . .	Water.	Water.	Water.	Water.	Water, 50-horse.	Water and Steam, 25-horse.
Number of stamps, . .	8.	6.	4.	4.	10.	10.
Kind of amalgamating machinery,	Copper plates.	Copper plates.	Copper plates.	Copper plates.	Copper plates; 1 Arrastra, 1 Beath's Grinder, 1 Salmon's Concentrator.	Copper plates.
Cost of roads or trails,	$1000.	$1500.	$50.00.	None.	$300.	$500.
Cost of flumes & ditches,	$1000.	$500.	$500.	$500.	$1200.	?

Number,	*25.*	*26.*	*27.*	*28.*	*29.*	*30.*
Mine,	**Morse's.**	**Orcutt's.**	**(No Mine.)**	**Eureka.**	**Summer's.**	**Grizzly.**
Location,	N'r S.Fork, Stanislaus.	N'r S.Fork, Stanislaus.	. . . . .	Summerville.	Near Summerville.	N. Fork, Tuolumne.
County,	Tuolumne.	Tuolumne.	Tuolumne.	Tuolumne.	Tuolumne.	Tuolumne.
Date of visit,	September 13th.	September 13th.	September 13th.	September 12th.	September 12th.	September 12th.
Whether worked,	Worked.	Worked.	. . . . .	Worked.	Worked.	Not worked.
Vein inclosed in,	Granite.	Granite.	. . . . .	Granite & Mica Slate.	Mica Slate.	Mica Slate.
Strike,	N. 30° E.	N. 50° E.	. . . . .	N. 25° W.	N. 25° W.	N.N.W.–S.S.E.
Dip,	E.S.E., 60°.	S.E., 70°.	. . . . .	E.N.E., 55°.	E.N.E., 75°.	E.N.E.
Average width,	1 foot 6 inches.	2 feet 6 inches.	. . . . .	4 feet.	3 feet.	7 feet.
Veinstone,	Bluish Quartz.	Bluish Quartz.	. . . . .	White Quartz.	Quartz.	Quartz.
Sulphurets present,	Iron, Lead.	Iron, Lead.	. . . . .	Iron.	Iron.	?
Yield of ore per ton,	$25 to $30.	$25.00.	. . . . .	$14.00.	$15.00.	$15 to $20.
Cost of extraction,	$2.00	$2.00.	. . . . .	$4 00.	$2.50.	$4.00.
Transportation to mill,	$0.50.	$0.50.	. . . . .	None.	$0.50.	$0.50.
Cost of treatment at mill,	$6.00.	$6.00.	. . . . .	$2.75.	$2.50.	$6.00.
Mill,	**(No Mill.)**	**Orcutt's.**	**Ryerson's.**	**Eureka.**	**Summer's.**	**Grizzly.**
Location,	. . . . .	Near Mine.	3 m. from William's Mt.	On Mine.	Summerville.	N. Fork, Tuolumne.
Whether running,	. . . . .	Running.	Running.	Running.	Not running.	Not running.
When erected,	. . . . .	August, 1864.	August, 1865.	May, 1855.	1860.	May, 1863.
Cost,	. . . . .	$3000.	$20,000.	$28,000.	$6000.	. . . . .
Power employed,	. . . . .	Water, 25-horse.	Water.	Water, 100-horse.	Water.	Water and steam.
Number of stamps,	. . . . .	5.	. . . . .	20.	8.	20.
Kind of amalgamating machinery,	. . . . .	Copper plates and Blankets.	1 Centrifugal Grinder, 1 Ryerson's Pulverizer, 1 Superheated Steam Apparatus, 1 Shaking Table.	4 Shaking pans, 1 Chili mill.	Copper plates; 1 Arrastra, 2 Shaking pans.	12 cast-iron barrels.
Cost of roads or trails,	$50.00.	$50.00.	. . . . .	. . . . .	$500.	$2200.
Cost of flumes & ditches,	None.	$100.	. . . . .	$1500.	$300.	$1200.

Number,	*31.*	*32.*	*33.*	*34.*	*35.*	*36.*
Mine,	**Excelsior.**	**Dagner.**	**Mount Vernon.**	**Monitor.**	**Green's.**	**Pirate.**
Location,	1 m. from Uniontown.	N. Fork, Tuolumne.	N. Fork, Tuolumne.	N. Fork, Tuolumne.	Near Uniontown.	N'r N. Fork, Tuolumne
County,	Tuolumne.	Tuolumne.	Tuolumne.	Tuolumne.	Tuolumne.	Tuolumne.
Date of visit,	September 14th.	September 14th.	September 14th.	September 14th.	September 14th.	September 14th.
Whether worked,	Worked.	Worked.	Worked.	Worked.	Worked.	Worked.
Vein inclosed in,	Granite.	Granite.	Granite.	Granite.	Granite.	Granite.
Strike,	N. 54° E.	N. 60° E.	N. 50° E.	N. 60° E.	N. 10° E.	N.E.–S.W.
Dip,	S.E., 35°.	S.E. 50°.	S.E., 45°	S.E., 55°.	E., 45°.	S.E., 45°.
Average width,	2 feet 6 inches.	1 foot.	1 foot.	1 foot 6 inches.	1 foot.	1 foot.
Veinstone,	Bluish Quartz.	Ferruginous Quartz.	White Quartz.	White Quartz.	White Quartz.	White Quartz.
Sulphurets present,	Iron, Lead, Zinc.	Iron, Lead.	Iron, Lead.	Iron, Lead.	?	Iron, Lead.
Yield of ore per ton,	$60.00.	$40.00.	?	$30.00.	$80.00.	$25.00.
Cost of extraction,	$3.00.	$4.50	?	$3.00.	$10.00.	$9.50.
Transportation to mill,	$0.50.	None.	?	None.	None.	$0.50.
Cost of treatment at mill,	$1.00.	$1.75.	?	$2.00.	$3.50.	$3.50.
Mill,	**Excelsior.**	**Dagner.**	**(No Mill.)**	**Monitor.**	**Green's.**	**Pirate.**
Location,	N. Fork, Tuolumne.	On Mine.	. . . . .	Near Mine.	On Mine.	N. Fork Tuolumne.
Whether running,	Not running.	Not running.	. . . . .	Running.	Running.	Running.
When erected,	1861.	November, 1864.	. . . . .	October, 1863.	March, 1865.	November, 1859.
Cost,	$13,500.	$22,000	. . . . .	$7000.	$4000.	$10,000.
Power employed,	Water, 20-horse.	Steam, 42-horse.	. . . . .	Water.	Steam, 15-horse.	Water.
Number of stamps,	10.	10.	. . . . .	5.	5.	10.
Kind of amalgamating machinery,	Copper plates and 1 Shaking pan.	Copper plates, 2 Arrastras, 1 Shaking table.	. . . . .	1 Ambler's Concentrator, 3 Arrastras, 1 Shaking table.	Copper plates, and 1 Beath's Amalgamator.	Copper plates, and 2 Arrastras.
Cost of roads or trails,	$2000.	None.	. . . . .	$1000.	None.	?
Cost of flumes & ditches,	$500.	None.	. . . . .	$300.	None.	?

Number,	*37.*	*38.*	*39.*	*40.*	*41.*	*42.*
Mine,	**Independence.**	**Great Eastern.**	**Comstock.**	**Soulsby.**	**Independent.**	**Gilson's (Old).**
Location,	1 mile from Excelsior.	2 m. from Williams's.	1 m. from Williams's.	Soulsbyville.	Near Soulsbyville.	Soulsbyville.
County,	Tuolumne.	Tuolumne.	Tuolumne.	Tuolumne.	Tuolumne.	Tuolumne.
Date of visit,	September 15th.	September 15th.	September 15th.	September 15th.	September 15th.	September 15th.
Whether worked,	Worked.	Worked.	Worked.	Worked.	Worked.	Not worked.
Vein inclosed in,	Granite.	Granite.	Granite.	Granite.	Granite.	Granite.
Strike,	N.E.–S.W.	N. 22° E.	N.–S.	N. 15° E.	N. 7° E.	N. 15° E.
Dip,	S.E., 45°.	?	E., 75°.	E., 80°.	E., 85°.	E., 85°.
Average width,	4 feet 6 inches.	2 feet.	1 foot 6 inches.	1 foot 3 inches.	10 inches.	1 foot 6 inches.
Veinstone,	Reddish Quartz.	Bluish and Reddish Quartz.	Reddish Quartz.	Quartz.	Ferruginous Quartz.	Quartz.
Sulphurets present,	Iron, Lead.	None seen.	Iron, Lead.	Iron, Lead, Zinc.	Iron, Lead.	Iron, Lead, Zinc.
Yield of ore per ton,	$40.00.	$20.00.	?	$25 to $30.	$36.00.	$52.50.
Cost of extraction,	$2.75 to $6.00.	?	$2.50.	$8.00.	?	$12.00.
Transportation to mill,	$0.50.	?	?	None.	?	None.
Cost of treatment at mill,	$2.75.	?	?	$3.75.	?	$4.50.
Mill,	**Independence.**	**(No Mill.)**	**(No Mill.)**	**Soulsby.**	**(No Mill.)**	**Gilson's.**
Location,	Near Mine.	. . . . .	. . . . .	On Mine.	. . . . .	On Mine.
Whether running,	Running.	. . . . .	. . . . .	Running.	. . . . .	Not running.
When erected,	1864.	. . . . .	. . . . .	1858.	. . . . .	1859.
Cost,	$15,000.	. . . . .	. . . . .	$20,000.	. . . . .	$9000.
Power employed,	Steam, 30-horse.	. . . . .	. . . . .	Steam, 35-horse.	. . . . .	Steam, 20-horse.
Number of stamps,	10.	. . . . .	. . . . .	20.	. . . . .	10.
Kind of amalgamating machinery,	Copper plates and blankets.	. . . . .	. . . . .	Copper plates.	. . . . .	Copper plates and 2 Arrastras.
Cost of roads or trails,	?	. . . . .	. . . . .	?	. . . . .	None.
Cost of flumes & ditches,	None.	. . . . .	. . . . .	None.	. . . . .	None.

Number,	43.	44.	45.	46.	47.	48.
Mine,	**Jackson.**	**Calder's.**	**(No Mine.)**	**Consuelo.**	**Water's.**	**Watt's.**
Location,	Williams's Ranch.	Calder's Ranch.	Calder's Ranch.	N. Fork, Tuolumne.	Jackass Hill.	N'r Robinson's Ferry.
County,	Tuolumne.	Tuolumne.	Tuolumne.	Tuolumne.	Tuolumne.	Tuolumne.
Date of visit,	September 16th.	September 16th.	September 16th.	September 22d.	September 28th.	September 28th.
Whether worked,	Worked.	Worked.		Worked.	Worked.	Worked.
Vein inclosed in,	Granite.	Granite.		Quartzo-micaceous Slate.	Green Schists.	Green Schists.
Strike,	N. 25° E.	N. 30° E.		N. 20° W.	N. 70° W.	N. 57° W.
Dip,	E.S.E., 45°.	S.E., 60°.		E.N.E., 50°.	N.N.E., 65°.	S.W., 50°.
Average width,	6 inches.	1 foot 2 inches.		3 feet 6 inches.	9 feet.	2 inches.
Veinstone,	Quartz.	Quartz.		White Quartz.	Quartz, Limestone, Slate.	Ferruginous Quartz.
Sulphurets present,	None seen.	Iron, Lead.		Iron, Lead.	Iron.	Iron, Copper.
Yield of ore per ton,	$107.	?		?	$6.00.	$180.
Cost of extraction,	?	?		?	$2.00.	$60.00.
Transportation to mill,	?	?		?	$0.25.	$0.25.
Cost of treatment at mill,	?	?		?	$1.75.	$7.00.
Mill,	**(No Mill.)**	**(No Mill.)**	**Wheeler's.**	**Consuelo.**	**Water's.**	**Watt's.**
Location,			Near Calder's.	Near Mine.	Near Mine.	Near Mine.
Whether running,			Not running.	Not finished.	Running.	Not running.
When erected,			?	November, 1865.	October, 1859.	April, 1862.
Cost,			$7500.	?	$3000.	$800.
Power employed,			Steam, 24-horse.	Water.	Water.	Water.
Number of stamps,			10.	20.	6.	3.
Kind of amalgamating machinery,			2 Varney's pans and 1 Concentrator.	Copper plates.	Copper plates.	1 Arrastra.
Cost of roads or trails,					$1000.	None.
Cost of flumes & ditches,					$600.	$25.00.

Number,	*49.*	*50.*	*51.*	*52.*	*53.*	*54.*
Mine,	**Union.**	**Alabama.**	**Gilson's (New).**	**(No Mine.)**	**Toledo.**	**Raw Hide.**
Location,	Near Tuttletown.	Whiskey Hill.	Curtis's Creek.	. . . . .	French Flat.	Raw Hide Ranch.
County,	Tuolumne.	Tuolumne.	Tuolumne.	Tuolumne.	Tuolumne.	Tuolumne.
Date of visit,	September 29th.	September 29th.	October 4th.	October 5th.	October 6th.	October 8th.
Whether worked,	Not worked.	Worked.	Worked.	. . . . .	Worked.	Worked.
Vein inclosed in,	Porphyritic Schists.	Talcose Slates.	Granite.	. . . . .	Talcose Slate.	Green Schist.
Strike,	N. 60° W.	N.N.W.–S.S.E.	N., 15° E.	. . . . .	N.W.–S.E.	. . . . .
Dip,	N.E., 70°.	E.N.E., 70°.	E., 80°.	. . . . .	N.E., 75°.	. . . . .
Average width,	2 feet.	25 feet.	1 foot 4 inches.	. . . . .	3 feet 6 inches.	4 feet.
Veinstone,	White Quartz.	Slate, Quartz.	Quartz.	. . . . .	Talcose Slate, Quartz.	White Quartz, Talc.
Sulphurets present,	Iron.	None seen.	Iron, Lead, Zinc.	. . . . .	Arsenic, Iron.	Iron, Lead.
Yield of ore per ton,	$8.00.	$10.00.	$40.00.	. . . . .	$10.00.	$25.00.
Cost of extraction,	$1.00.	$1.50.	$13.00.	. . . . .	$2.00.	?
Transportation to mill,	$0.25.	None.	$1.25.	. . . . .	None.	?
Cost of treatment at mill,	$1.75.	$1.00.	$6.00.	. . . . .	$2.00.	?
Mill,	**Union.**	**Alabama.**	**(Mill No. 42.)**	**Washington.**	**Labitour's.**	**Raw Hide.**
Location,	Mormon Creek.	On Mine.	. . . . .	E. of Bald Mountain.	On Mine.	Near Mine.
Whether running,	Not running.	Running.	. . . . .	Not running.	Not running.	Running.
When erected,	September, 1865.	November, 1858.	. . . . .	April, 1860.	January, 1863.	April, 1863.
Cost,	$2000.	$1500.	. . . . .	$1200.	$14,800.	?
Power employed,	Water.	Water.	. . . . .	Water.	Water.	Water.
Number of stamps,	8.	4.	. . . . .	3.	15.	10.
Kind of amalgamating machinery,	Copper plates.	Copper plates.	. . . . .	Copper plates.	Copper plates.	Copper plates.
Cost of roads or trails,	$50.00.	$150.	$1000.	None.	$1000.	None.
Cost of flumes & ditches,	$500.	$200.	None.	None.	$4000.	?

Number, ...	*55.*	*56.*	*57.*	*58.*	*59.*	*60.*
Mine, ...	**Shanghai.**	**Columbia.**	**Patterson's.**	**Valparaiso.**	**Turner's.**	**Preston's.**
Location, ...	Near Yankee Hill.	Near Columbia.	Near Tuttletown.	Near Tuttletown.	Turner's Flat.	Whiskey Hill.
County, ...	Tuolumne.	Tuolumne.	Tuolumne.	Tuolumne.	Tuolumne.	Tuolumne.
Date of visit, ...	October 10th.	October 11th.	October 13th.	October 13th.	October 17th.	October 20th.
Whether worked, . .	Worked.	Not worked.	Worked.	Not worked.	Worked.	Not worked.
Vein inclosed in, . .	Mica Slate.	Mica Slate.	Green Schists.	Mica Slate.	Porphyritic Schists.	Talcose Slate.
Strike, ...	N.N.W.–S.S.E.	N.–S.	N. 60° W.	N.–S.	N. 65° E.	N. 55° W.
Dip, ...	E.N.E., 50°–55°.	E., 40°.	N.E., 65°.	W., 85°.	N.N.W., 45°–60°.	N.E., 75°.
Average width, . . .	2 feet 6 inches.	3 feet 6 inches.	4 feet.	2 feet.	2 feet 6 inches.	10 inches.
Veinstone, ...	Ferruginous Quartz.	White Quartz.	White Quartz.	White Quartz.	Pure White Quartz.	White Quartz.
Sulphurets present, .	Iron, Copper.	Iron, Lead.	Iron, in crystals.	Iron, in crystals.	None seen.	Iron, Copper, Lead.
Yield of ore per ton, .	$40.00.	$9.00	$8.00.	$80.00.	Coarse Gold.	?
Cost of extraction, . .	$5.00.	$2.00.	$2.00.	?	?	$2.00.
Transportation to mill,	None.	$0.60.	$0.25.	None.	None.	$0.60.
Cost of treatment at mill,	$2.00.	$2.50.	$2.50.	?	None.	$1.50.
Mill, ...	**Shanghai.**	**Columbia.**	**Patterson's.**	**Valparaiso.**	**(No Mill.)**	**Preston's.**
Location, ...	On Mine.	Near Mine.	Near Mine.	On Mine.	. . . . .	Wood's Creek.
Whether running, . .	Running.	Not running.	Not running.	In ruins.	. . . . .	Not running.
When erected, . . .	June, 1865.	May, 1864.	May, 1858.	?	. . . . .	September, 1862.
Cost, ...	$4000.	$15,000.	$6500.	$3000.	. . . . .	$4000.
Power employed, . .	Water.	Water, 96-horse.	Water.	Water.	. . . . .	Water.
Number of stamps, . .	10.	15.	10.	6.	. . . . .	10.
Kind of amalgamating machinery,	Copper plates.	Copper plates.	Copper plates.	Copper plates.	. . . . .	Copper plates.
Cost of roads or trails,	None.	$1000.	$400.	None.	. . . . .	$300.
Cost of flumes & ditches,	$100.	$4000.	$1200.	?	. . . . .	$500.

Number,	*61.*	*62.*	*63.*	*64.*	*65.*	*66.*
Mine,	**Italian.**	**Old Whiskey Hill.**	**Nyman's.**	**John Knox's.**	**(No Mine.)**	**Clio.**
Location,	Whiskey Hill.	Whiskey Hill.	Near Quartz Mount'n.	Quartz Mountain.	. . . . .	Near Jacksonville.
County,	Tuolumne.	Tuolumne.	Tuolumne.	Tuolumne.	Tuolumne.	Tuolumne.
Date of visit,	October 20th.	October 20th.	October 20th.	October 20th.	October 20th.	October 25th.
Whether worked,	Not worked.	Worked.	Worked.	Worked.	. . . . .	Worked.
Vein inclosed in,	Talcose Slates.	Talcose Slates.	Talcose Slates.	Talcose Slates.	. . . . .	Schists & Serpentine.
Strike,	N. 50° W.	N. 50°. W.	N. 50° W.	N.W.–S.E.	. . . . .	N. 80° W.
Dip,	N.E., 80°.	N.E., 85°.	N.E., 50°.	N.E., 60°.	. . . . .	N., 60°.
Average width,	6 inches.	15 feet.	4 feet.	1 foot.	. . . . .	5 feet 6 inches.
Veinstone,	White Quartz.	Talcose Slate.	White Quartz.	Quartz.	. . . . .	White Quartz.
Sulphurets present,	Iron, Copper.	Copper, Iron.	Iron, Copper.	None seen.	. . . . .	Iron, in crystals.
Yield of ore per ton,	$20.00.	$15.00.	$17.00.	$65.00.	. . . . .	$15.00.
Cost of extraction,	$4.00.	$0.50.	$3.00.	?	. . . . .	$3.00.
Transportation to mill,	None.	$0.50.	$0.90.	?	. . . . .	$0.75.
Cost of treatment at mill,	$1.50.	$1.50.	$1.00.	?	. . . . .	$2.00.
Mill,	**Occidental.**	**Wood's Crossing.**	**Nyman's.**	**(No Mill.)**	**Widow Hill.**	**Clio.**
Location,	On Mine.	Wood's Creek.	Wood's Creek.	. . . . .	E. of Quartz Mountain.	Jacksonville.
Whether running,	Not running.	Running.	Running.	. . . . .	Running.	Not running.
When erected,	1857.	1857.	1862.	. . . . .	1861.	1858.
Cost,	$6000.	$2000.	$4000.	. . . . .	?	$5000.
Power employed,	Water.	Water.	Water.	. . . . .	Water.	Water.
Number of stamps,	12.	4.	10.	. . . . .	5.	10.
Kind of amalgamating machinery,	Copper plates and 1 Arrastra.	Copper plates.	Copper plates and 2 Arrastras.	. . . . .	Copper plates.	Copper plates and 2 Knox's pans.
Cost of roads or trails,	$400.	$1000.	$1000.	. . . . .	None.	$1000.
Cost of flumes & ditches,	$300.	$1200.	$1000.	. . . . .	?	$16,000.

Number, ...	67.	68.	69.	70.	71.	72.
Mine, ...	**Shawmut.**	**Josephine.**	**Eagle.**	**Italian.**	**Nonpareille.**	**Burns's.**
Location, ...	Blue Gulch.	Blue Gulch.	Blue Gulch.	1 mile from Deer Flat.	Deer Flat.	First Garrote.
County, ...	Tuolumne.	Tuolumne.	Tuolumne.	Tuolumne.	Tuolumne.	Tuolumne.
Date of visit, ...	October 25th.	October 25th.	October 25th.	October 27th.	October 27th.	October 28th.
Whether worked, . .	Worked.	Worked.	Worked.	Worked.	Not worked.	Worked.
Vein inclosed in, . .	Great Quartz vein & Talcose Slate.	Great Quartz vein & Talcose Slate.	Great Quartz vein & Talcose Slate.	Mica Slate.	Mica Slate.	Mica Slate.
Strike, ...	N. 50° W.	N. 50° W.	N.W.–S.E.	E.–W.	E.–W.	N. 60° E.
Dip, ...	. . . . .	. . . . .	N.E., 60°.	S., 30°.	N. 60°.	N.N.W., 70°.
Average width, . . .	1 foot 6 inches.	8 feet.	2 feet.	8 inches.	?	4 feet.
Veinstone, ...	Ribboned Quartz.	White Quartz.	White Quartz, Slates.	Ferruginous Quartz.	White Quartz, Slates.	Slates, Quartz.
Sulphurets present, .	Iron, Lead.	Iron, crystallized.	Iron, Lead.	Iron, Lead.	Iron.	None seen.
Yield of ore per ton, .	$25.00.	$8.00.	$12.00.	Coarse Gold.	$30.00.	$20.00.
Cost of extraction, . .	$2.75.	$2.00.	$3.50.	?	?	$1.00.
Transportation to mill,	None.	None.	$0.50	?	?	$1.12½.
Cost of treatment at mill,	$1.50.	$1.50.	$2.00.	?	?	$4.50.
Mill, ...	**Shawmut.**	**Stetson's.**	**Eagle.**	**(No Mill.)**	**Duprat's.**	**(No Mill.)**
Location, ...	Near Mine.	Wood's Creek.	Blue Gulch.	. . . .	Near Mine.	. . . .
Whether running, . .	Not running.	Not running.	Running.	. . . .	Not running.	. . . .
When erected, . . .	January, 1859.	October, 1865.	1861.	. . . .	June, 1864.	. . . .
Cost, ...	$8000.	$7500.	$11,000.	. . . .	?	. . . .
Power employed, . .	Steam, 25-horse.	Water, 30-horse.	Water.	. . . .	Water (turbine).	. . . .
Number of stamps, . .	10.	. . . . .	10.	. . . .	5.	. . . .
Kind of amalgamating machinery,	Copper plates and 4 Knox's pans.	Copper plates, 5 wooden pans, 1 Brodie's Cracker, 1 Centrifugal Pulverizer, and 1 Grinder.	Copper plates.	. . . .	Copper plates, 1 Farrand's Amalgamator, 1 Settler.	. . . .
Cost of roads or trails,	$2000	$800.	$2500.	. . . .	?	. . . .
Cost of flumes & ditches,	None.	$500.	$3000.	. . . .	?	. . . .

Number,	73.	74.	75.	76.	77.	78.
Mine,	**(No Mine.)**	**Second Garrote.**	**Morkam.**	**Kanaka.**	**Phœnix.**	**Mohrmann's.**
Location,	. . . .	Big Creek.	Big Creek.	Kanaka Creek.	Big Oak Flat.	Near Big Oak Flat.
County,	Tuolumne.	Tuolumne.	Tuolumne.	Tuolumne.	Tuolumne.	Tuolumne.
Date of visit,	October 28th.	October 28th.	October 28th.	October 28th.	October 31st.	November 1st.
Whether worked,	. . . .	Worked.	Not worked.	Not worked.	Worked.	Worked.
Vein inclosed in,	. . . .	Mica Slate.	Mica Slate.	Mica Slate.	Mica Slate & Granite.	Mica Slate & Granite.
Strike,	. . . .	E.–W.	N. 55° E.	N.–S.	W.N.W.–E.S.E.	W.N.W.–E.S.E.
Dip,	. . . .	N. 65°.	N.W., 65°.	E., 55°.	N.E., 20°–30°.	N.E., 22°–25°.
Average width,	. . . .	1 foot 6 inches.	4 feet.	?	2 feet 6 inches.	2 feet 6 inches.
Veinstone,	. . . .	Quartz.	White Quartz.	White Quartz.	White Quartz.	White Quartz.
Sulphurets present,	. . . .	Iron.	Iron, Lead, Zinc.	Iron.	Iron, Lead.	Iron, Lead.
Yield of ore per ton,	. . . .	?	$14.00.	$8.00.	$15.00.	?
Cost of extraction,	. . . .	?	$1.50.	$2.00.	?	?
Transportation to mill,	. . . .	?	$0.40.	$2.00.	?	?
Cost of treatment at mill,	. . . .	?	$3.00.	$3.00.	?	?
Mill,	**Cross's.**	**Mill No. 75.**	**Pacific.**	**Mill No. 75.**	**Phœnix.**	**(No Mill.)**
Location,	Near Big Oak Flat.	. . . .	Big Creek.	. . . .	Near Mine.	. . . .
Whether running,	Running.	. . . .	Not running.	. . . .	Not running.	. . . .
When erected,	1863.	. . . .	1863.	. . . .	?	. . . .
Cost,	$4500.	. . . .	$3000.	. . . .	$200.	. . . .
Power employed,	Water.	. . . .	Water.	. . . .	Water.	. . . .
Number of stamps,	10.	. . . .	5.	. . . .	1 Arrastra.	. . . .
Kind of amalgamating machinery,	3 Varney's Pans & 1 Settler.	. . . .	Copper plates and 1 Arrastra.	. . . .	The same Arrastra.	. . . .
Cost of roads or trails,	. . . .	. . . .	?	. . . .	None.	. . . .
Cost of flumes & ditches,	?	. . . .	?	. . . .	?	. . . .

Number,	*79.*	*80.*	*81.*	*82.*	*83.*	*84.*
Mine,	**Kenney's.**	**Golden Rule.**	**Mine No. 80.**	**Mine No. 80.**	**Brown's Flat.**	**Zuckermann's.**
Location,	Big Oak Flat.	Near Poverty Hill.	. . . .	. . . .	Brown's Flat.	Near Brown's Flat.
County,	Tuolumne.	Tuolumne.	Tuolumne.	Tuolumne.	Tuolumne.	Tuolumne.
Date of visit,	November 1st.	November 3d.	November 3d.	October 28th.	November 6th.	November 6th.
Whether worked,	Worked.	Worked.	. . . .	. . . .	Not worked.	Worked.
Vein inclosed in,	Mica Slate & Granite.	Porphyritic Schists & Great Quartz vein.	. . . .	. . . .	Mica Slate.	Mica Slate.
Strike,	W.N.W.–E.S.E.	N. 30° W.	. . . .	. . . .	N.–S.	N.–S.
Dip,	N., 30°–35°.	N.E., 70°–75°.	. . . .	. . . .	W., 45°.	E. 40°.
Average width,	5 feet.	8 feet.	. . . .	. . . .	6 inches.	1 foot 6 inches.
Veinstone,	White Quartz.	Quartz and Slate.	. . . .	. . . .	White Quartz.	White Quartz.
Sulphurets present,	Iron, Lead.	Iron.	. . . .	. . . .	None seen.	None seen.
Yield of ore per ton,	$15.00.	$40.00.	. . . .	. . . .	$40.00.	$40.00.
Cost of extraction,	?	$1.00	. . . .	. . . .	$11.00.	$5.00.
Transportation to mill,	?	?	. . . .	. . . .	None.	?
Cost of treatment at mill,	?	$2.00.	. . . .	. . . .	$1.50.	$4.00.
Mill,	**Kenney's.**	**Golden Rule No. 1.**	**Golden Rule No. 2.**	**Golden Rule No. 3.**	**Brown's Flat.**	**Zuckermann's.**
Location,	On Mine.	Sullivan's Creek.	Near Mine.	On Mine.	On Mine.	Near Mine.
Whether running,	Not running.	Running.	Not running.	Not finished.	Not running.	Not running.
When erected,	1860.	1860.	1860.	December, 1865.	1856.	1859.
Cost,	$1000.	$2000.	$1300.	$12,000.	$3500.	$4000.
Power employed,	Water.	Water.	Water.	Water.	Water.	Water.
Number of stamps,	2 large Arrastras.	5.	2.	15.	4.	4.
Kind of amalgamating machinery,	The same Arrastras.	Copper plates.	2 Arrastras.	Copper plates, 1 Arrastra, 3 Shaking tables.	Copper plates, and 1 Arrastra.	Copper plates.
Cost of roads or trails,	None.	$50.00.	. . . .	None.	None.	?
Cost of flumes & ditches,	?	$400.	$200.	?	$1500.	?

www.ingramcontent.com/pod-product-compliance
Lightning Source LLC
LaVergne TN
LVHW020641110826
845149LV00004B/1306

* 9 7 8 1 4 1 8 1 9 0 4 0 8 *